Measuring - Grade 2

A complete workbook with lessons and problems

By Maria Miller

ISBN 978-1522843610

EDITION 7/2016

Contents

Preface

Hello! I am Maria Miller, the author of this math book. I love math, and I also love teaching. I hope that I can help you to love math also!

I was born in Finland, where I also grew up and received all of my education, including a Master's degree in mathematics. After I left Finland, I started tutoring some home-schooled children in mathematics. That was what sparked me to start writing math books in 2002, and I have kept on going ever since.

In my spare time, I enjoy swimming, bicycling, playing the piano, reading, and helping out with Inspire4.com website. You can learn more about me and about my other books at the website MathMammoth.com.

This book, along with all of my books, focuses on the conceptual side of math... also called the "why" of math. It is a part of a series of workbooks that covers all math concepts and topics for grades 1-7. Each book contains both instruction and exercises, so is actually better termed *worktext* (a textbook and workbook combined).

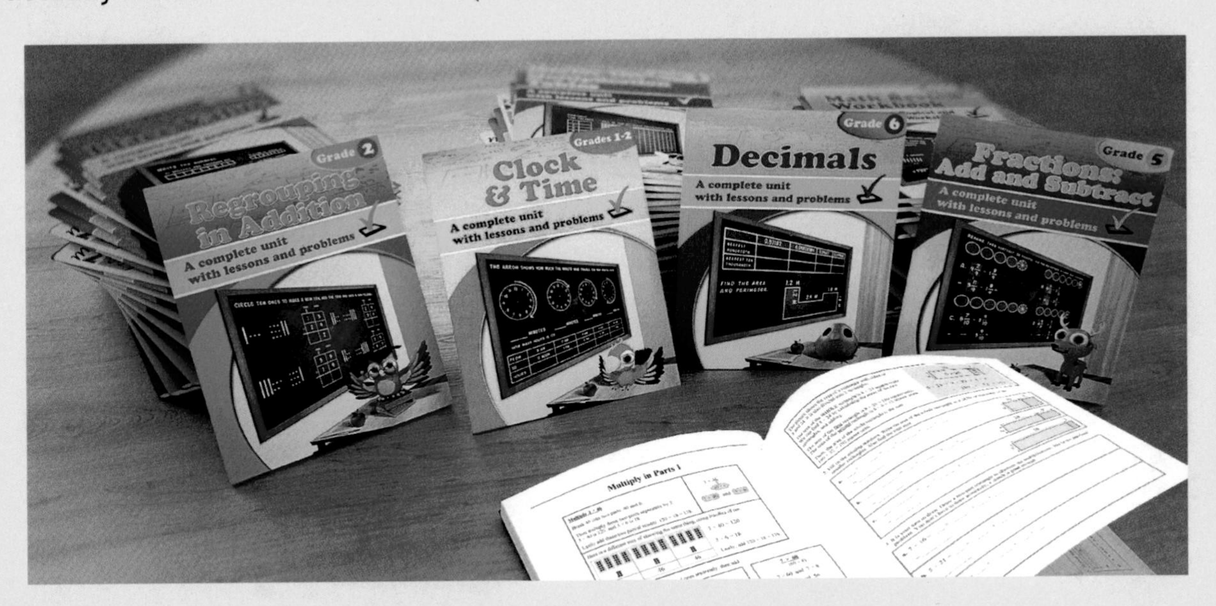

My lower level books (approximately grades 1-5) explain a lot of mental math strategies, which help build number sense – proven in studies to predict a student's further success in algebra.

All of the books employ visual models and exercises based on visual models, which, again, help you comprehend the "why" of math. The "how" of math, or procedures and algorithms, are not forgotten either. In these books, you will find plenty of varying exercises which will help you look at the ideas of math from several different angles.

I hope you will enjoy learning math with me!

Introduction

Measuring - Grade 2 is a complete workbook with lessons and exercises dealing with measuring length and weight. The student measures and estimates length in inches and half-inches, and learns to measure to the nearest half-inch or to the nearest centimeter. The bigger units—feet, miles, meters, and kilometers—are introduced, but in this grade level the students do not yet study conversions between the units.

The lessons on measuring weight have several activities to do at home using a bathroom scales. The goal is to let students become familiar with pounds and kilograms, and have an idea of how many pounds or kilograms some common things weigh.

When it comes to measuring, experience is the best teacher. So, encourage your child to use measuring devices (such as a measuring tape, ruler, and scales), and to "play" with them. In this way, the various measuring units start to become a normal part of his/her life, and will never be forgotten.

I wish you success in math teaching!

Maria Miller, the author

Helpful Resources on the Internet

Use these free online resources to supplement the "bookwork" as you see fit.

Measure It!
Click on the ruler to measure a red bar.
http://onlineintervention.funbrain.com/measure/index.html

Reading a tape measure worksheets
Worksheet generator: You can choose to what accuracy to measure, inches, or inches and feet.
http://themathworksheetsite.com/read_tape.html

Inchy Picnic Game
Measure with a ruler to find how many inches Andy Ant needs to go.
http://www.fuelthebrain.com/games/inchy-picnic/

Measuring activity
Measure the given lines with a centimeter-ruler, including lines you draw on your own.
http://www.taw.org.uk/lic/itp/itps/ruler_1_2.swf

Measurement Quiz
Practice measuring fish in centimeters.
http://www.thatquiz.org/tq-9/math/measurement/

US Standard Measurements for Length
This page has clear explanations and good illustrations of the standard units for measuring length.
https://www.mathsisfun.com/measure/us-standard-length.html

Measuring
Choose to measure with a metric or an imperial ruler and choose a level of difficulty.
http://www.abcya.com/measuring.htm

Reading Scales
Helps teachers to illustrate a variety of measuring devices and how to read them.
http://www.teacherled.com/2009/02/18/reading-scales-2/

Scales Reader
Practice reading the scales in grams and/or kilograms.
http://www.ictgames.com/weight.html

Measuring scales
An interactive scales for the purpose of demonstrating how a scales works. You can add weights to the scales and choose to show or hide the total weight.
http://www.taw.org.uk/lic/itp/itps/measuringScales_1_8.swf

Mostly Postie!
Choose "kg and half kg". Place a package on the scale, and enter the reading, including the possible "1/2 kg."
http://www.ictgames.com/mostlyPostie.html

Measuring to the Nearest Centimeter

Remember? We can measure how long things are using *centimeters*.

This line is 1 centimeter long: ⊢⊣

A centimeter is written in short form as "cm."

The blue line on the right is 6 cm long. →

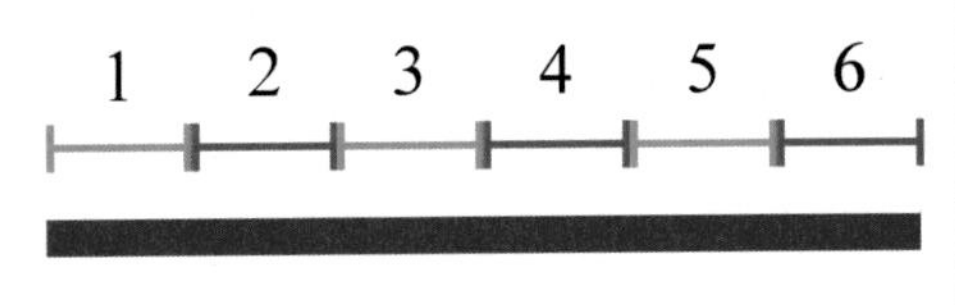

1. How many centimeters long are these lines?

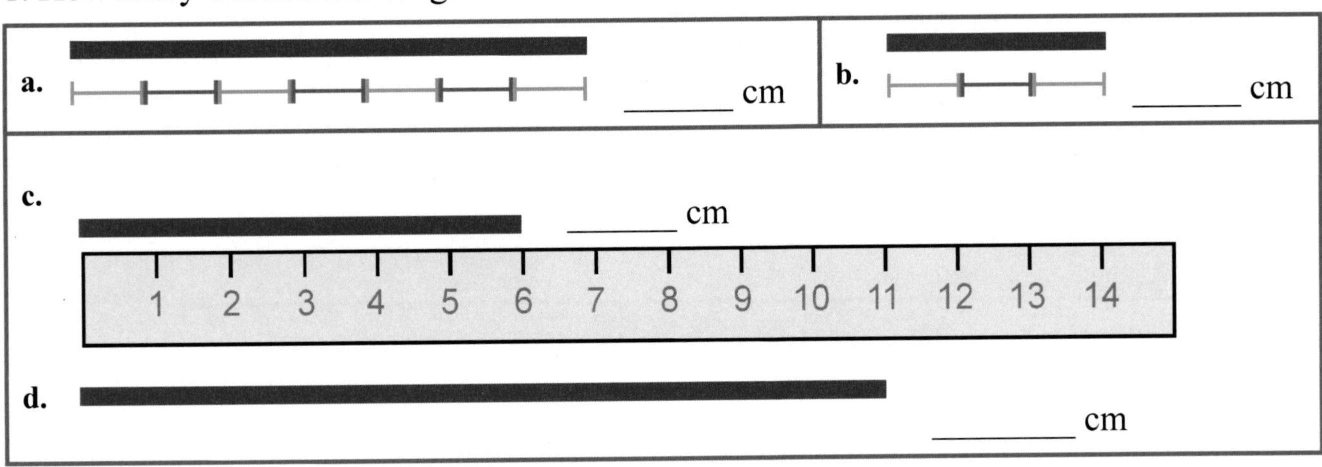

2. Measure the pencils with a centimeter ruler. If you don't have one, you can cut out the one from the bottom of this page. Then answer the questions.

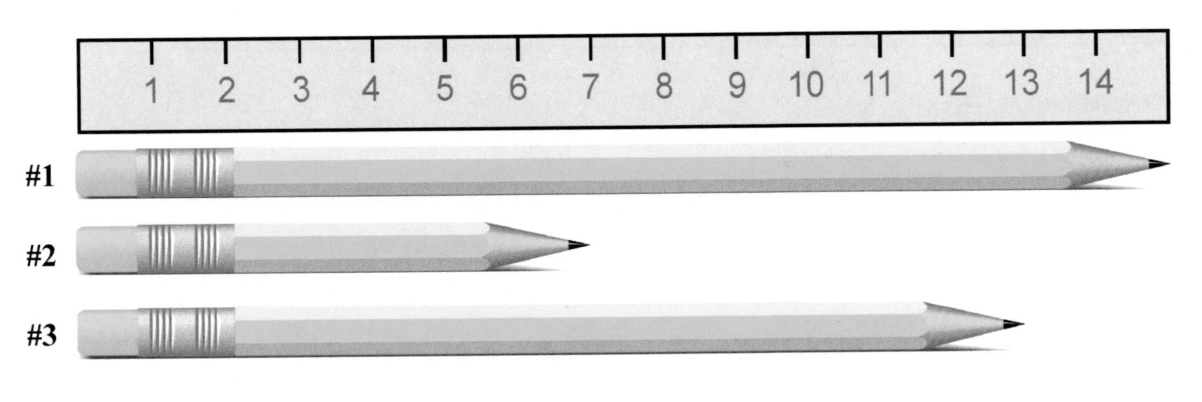

a. How much longer is pencil #1 than pencil #2? _______ cm

b. How much longer is pencil #3 than pencil #2? _______ cm

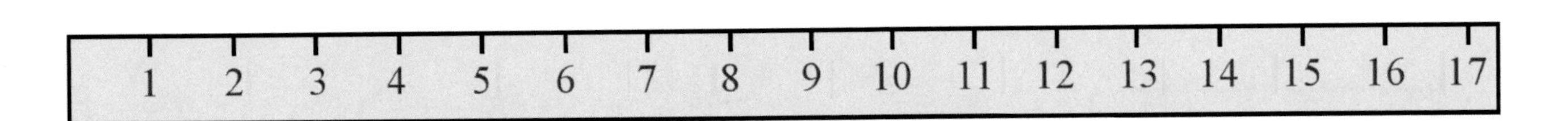

Most things are NOT exactly a certain number of whole centimeters. You can measure them to the nearest centimeter.

The pencil below is a little over 10 cm long. It is *about 10* cm long.

1 2 3 4 5 6 7 8 9 10 11 12 13 14

This pencil is about 9 cm long. The end of the pencil is closer to 9 cm than to 8 cm.

1 2 3 4 5 6 7 8 9 10 11 12 13 14

3. Circle the number that is nearest to each arrow.

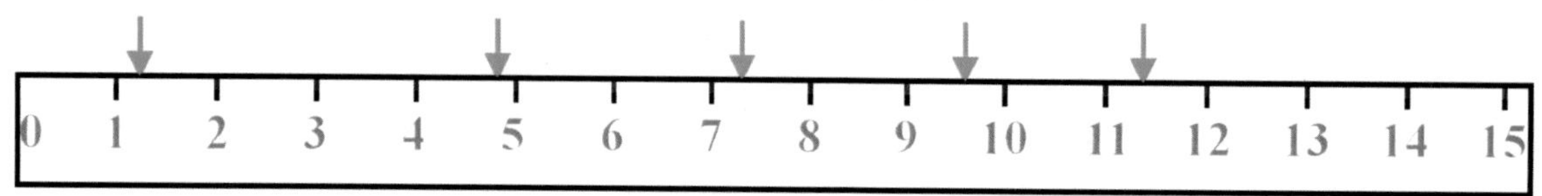

4. Measure the lines to the nearest centimeter.

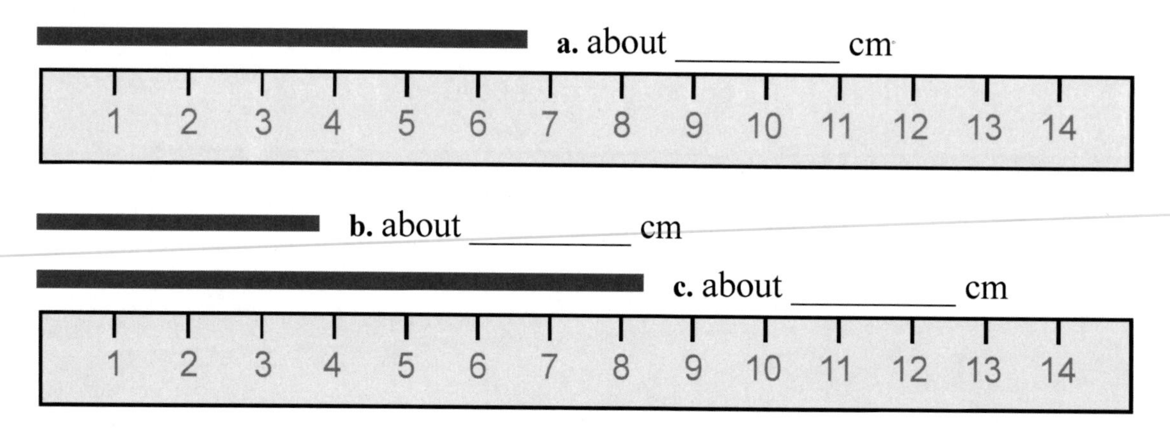

a. about _________ cm

b. about _________ cm

c. about _________ cm

5. This line is 1 cm long: |——| . Your finger is probably about that wide; put it on top of the 1-cm line and check! Guess how long these lines are. Then measure.

	My guess:	**Measurement:**
a. ________________	about _______ cm	about _______ cm
b. ______	about _______ cm	about _______ cm
c. __________	about _______ cm	about _______ cm

6. **a.** Find two small objects. Measure to find *about* how many centimeters longer one is than the other.

The ________________________________ is *about* _________ cm longer

than the ________________________________.

b. Find two other small objects. Measure to find *about* how many centimeters longer one is than the other.

The ________________________________ is *about* _________ cm longer

than the ________________________________.

7. Draw some lines here or on blank paper. Use a ruler. Hold the ruler down tight with one hand, while drawing the line with the other. It takes some practice!

a. 6 cm long

b. 3 cm long

c. 12 cm long

d. 17 cm long

8. Find some small objects. First GUESS how long or tall they are. Then measure. If the item is not exactly so-many centimeters long, then measure it to the nearest centimeter and write "about" before your cm-number, such as *about 8 cm.*

Item	GUESS	MEASUREMENT
	cm	cm
	cm	cm
	cm	cm
	cm	cm
	cm	cm

Inches and Half-Inches

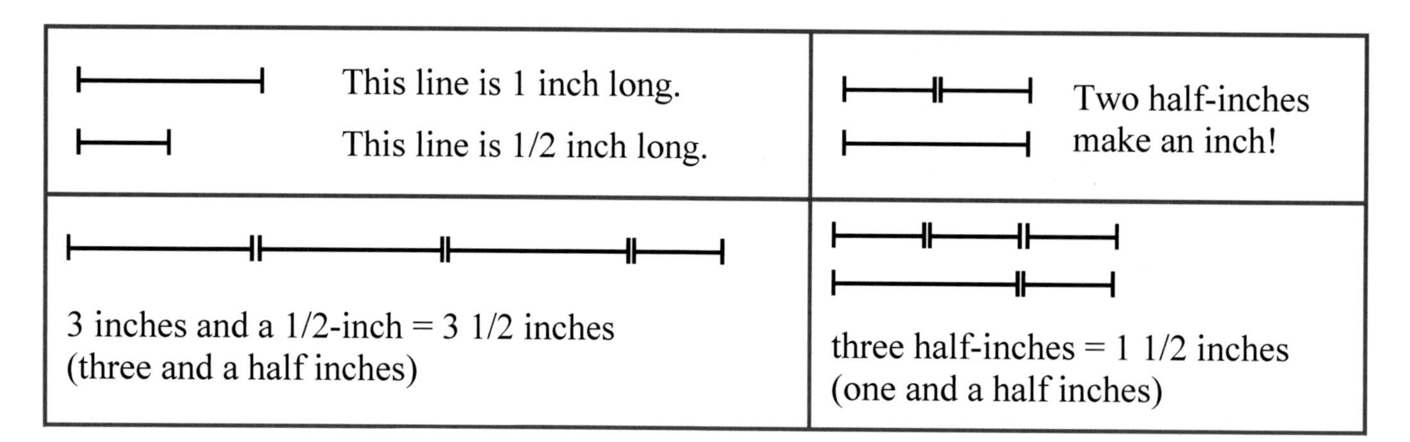

1. How long are the lines of inches and half-inches when placed end-to-end?

a. ________ inches

b. ________ inches

c. ________ inches

d. ________ in.

2. How long are these things in inches?

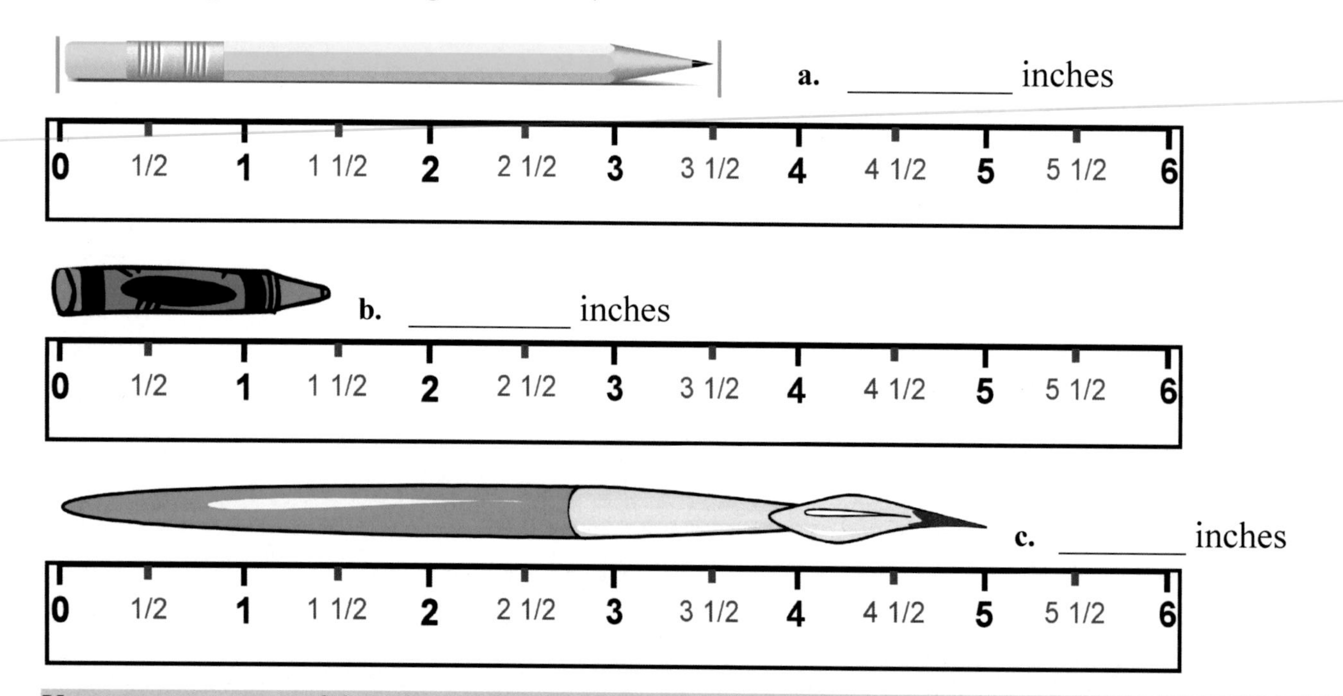

You can cut out one of the rulers in this lesson and tape it on an existing ruler or cardboard after you have finished the exercises on this and the next page!

Most objects are NOT exactly a certain number of whole inches, or even whole and half inches. You can measure them to the nearest inch, or to the nearest half-inch.

The pencil below is a little over 4 inches long. It is <u>*about* 4 inches long</u>.

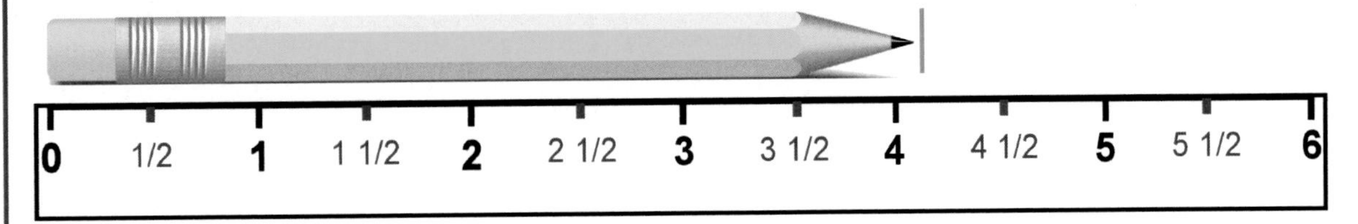

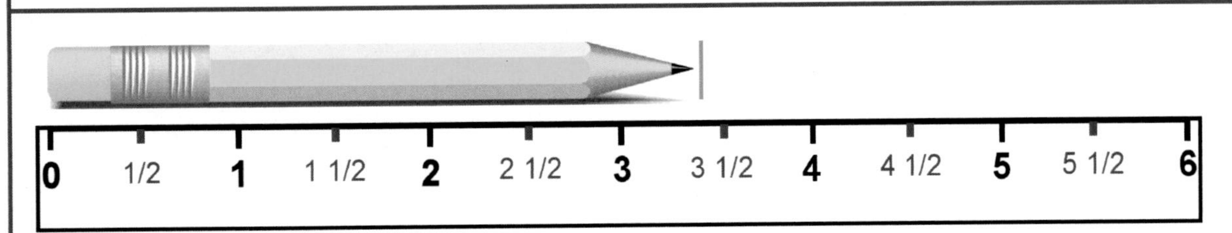

The pencil above is about 3 1/2 inches long. The end of the pencil is closer to 3 1/2 than to 3.

3. Circle the whole-inch or half-inch number that is nearest to each arrow.

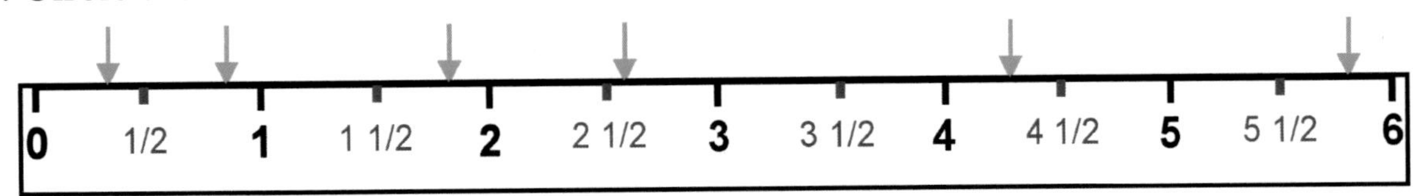

4. Measure the pencils to the nearest half-inch.

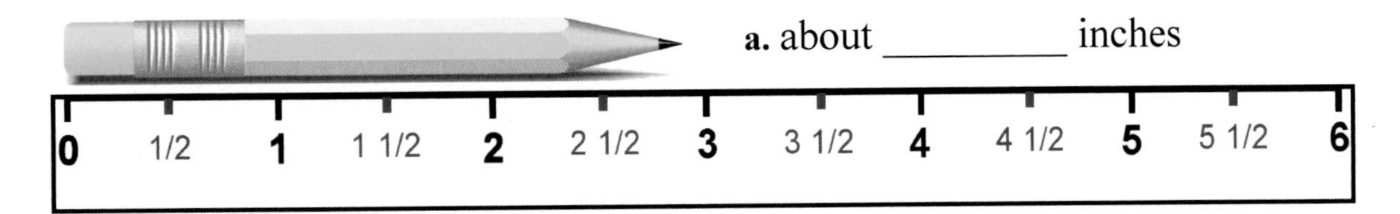

a. about _________ inches

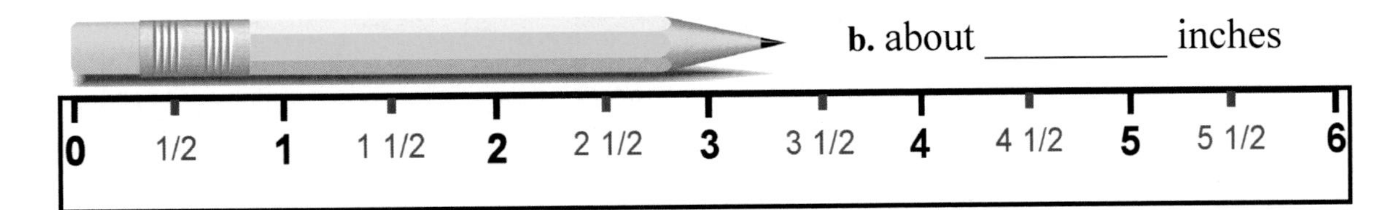

b. about _________ inches

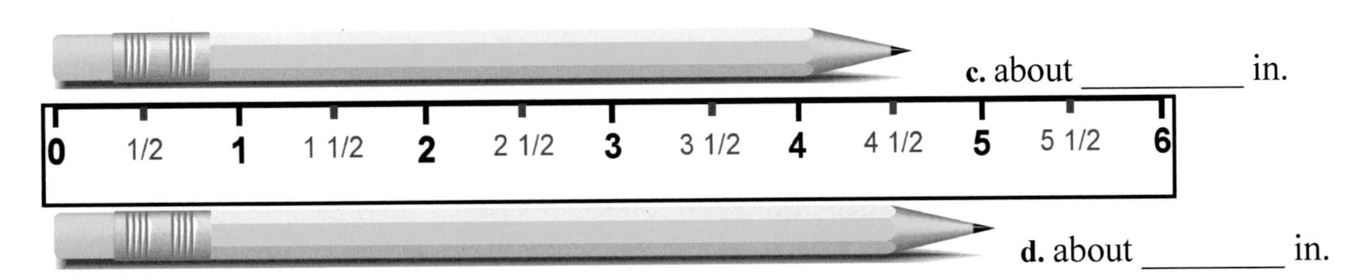

c. about _________ in.

d. about ________ in.

5. First GUESS how long these lines are in inches and half-inches. Write down your guess. After that, measure how long the lines are.

	GUESS	MEASUREMENT
a.	________ inches	________ inches
b.	________ inches	________ inches
c.	________ inches	________ inches

6. Draw some lines on a blank paper. Use a ruler. Hold the ruler down tight with one hand, while drawing the line with the other. It takes some practice!

a. 5 in. long **b.** 2 in. long

c. 12 in. long **d.** 9 in. long

7. Write the names of these shapes. Measure the sides of the shapes. "All the way around" means you need to find the *total* length of the four sides (use addition!).

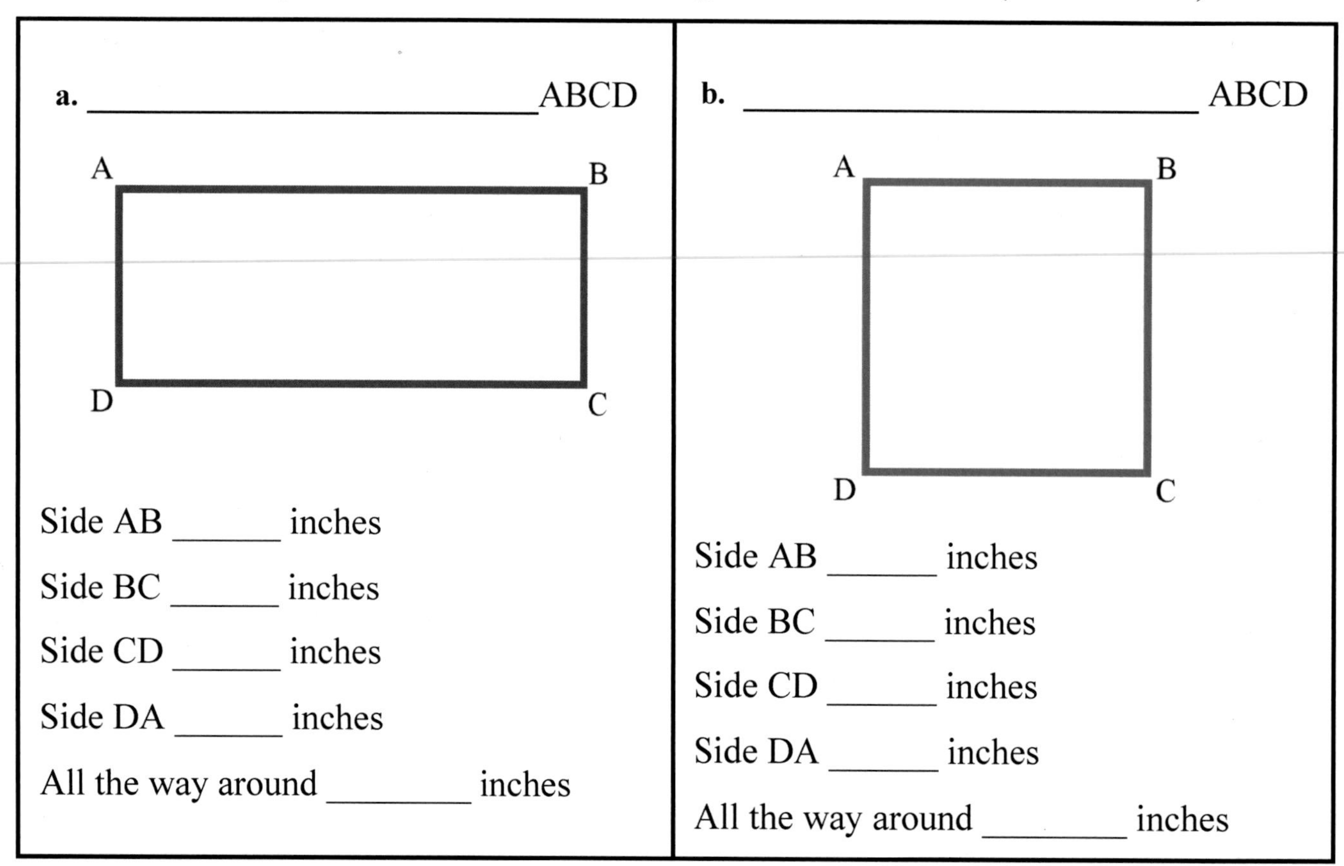

Some More Measuring

1. Jackie measured the length of a bunch of pencils at her home. She recorded her results in a line plot below. For each pencil, she put an "x" mark above the number line, to show how many centimeters long it was.

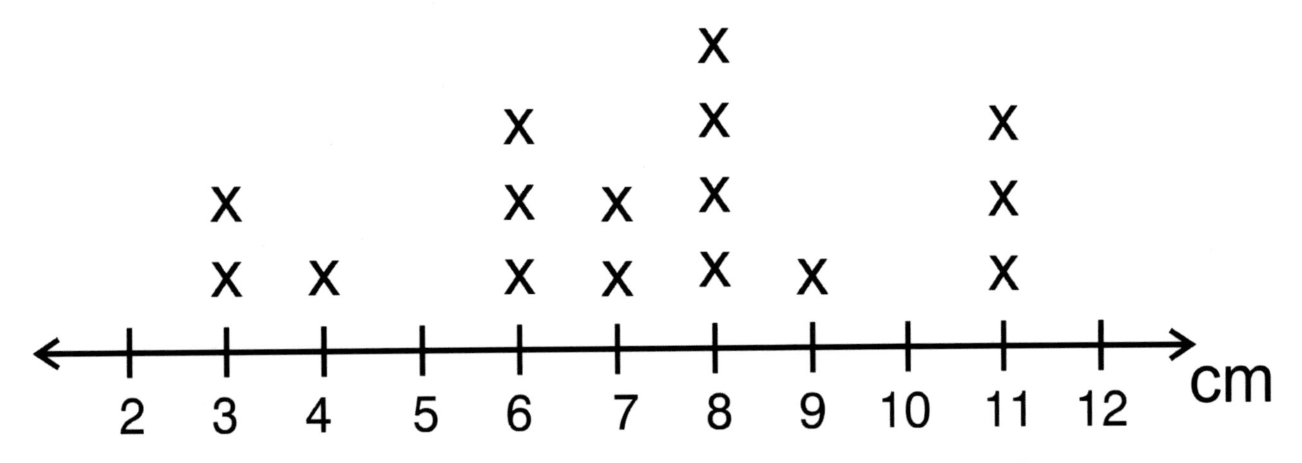

a. How many of Jackie's pencils were 3 cm long?

b. How many were 8 cm long?

c. How many pencils were 9 cm or longer?

d. How many pencils were 5 cm or shorter?

e. How long is Jackie's longest pencil? Her shortest pencil?

How much longer is the longest pencil than the shortest pencil?

2. Join these dots with lines to form a four-sided shape. What is the name for the shape?

____

Measure its sides to the nearest centimeter. Write "about ___ cm" next to each side.

How many centimeters is the *perimeter?*

(all the way around the shape) It is ________ cm.

3. Measure many pencils of different lengths to the *nearest* whole centimeter. Write the lengths below. (You don't have to measure as many pencils as there are empty lines.)

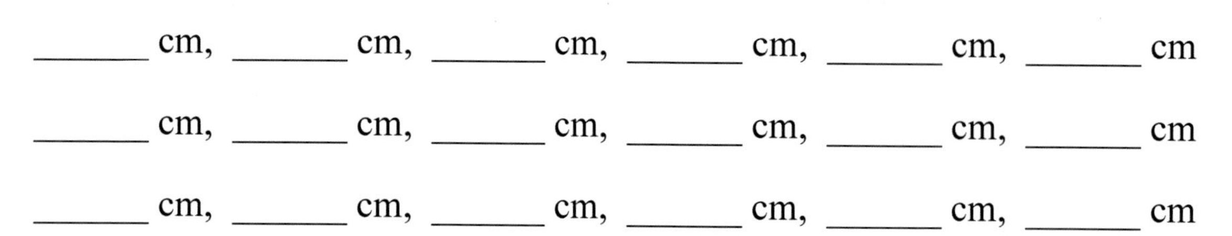

Now, make a line plot about your pencils like what Jackie made. Write an "X" mark for each pencil.

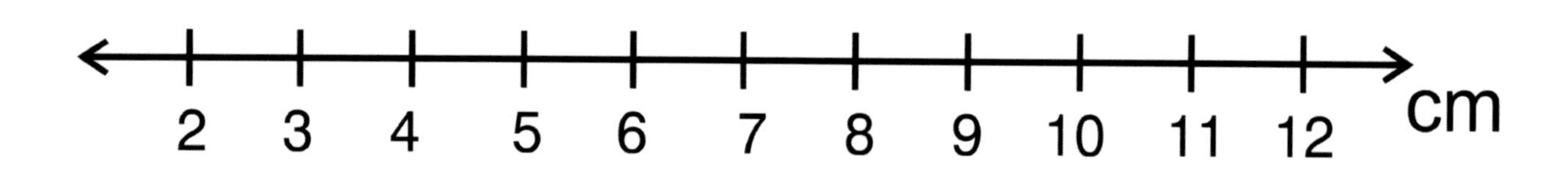

a. How much longer is your longest pencil than your shortest pencil?

b. If you take your three longest pencils and put them end-to-end, how long is your line of pencils? Add to find out.

It is ____________ cm. (If you can, measure to check your answer.)

4. Measure all the sides of this triangle to the nearest half-inch. Also, find the *perimeter* (all the way around the triangle).

Side AB ________ in.

Side BC ________ in.

Side CA ________ in.

Perimeter ________ in.

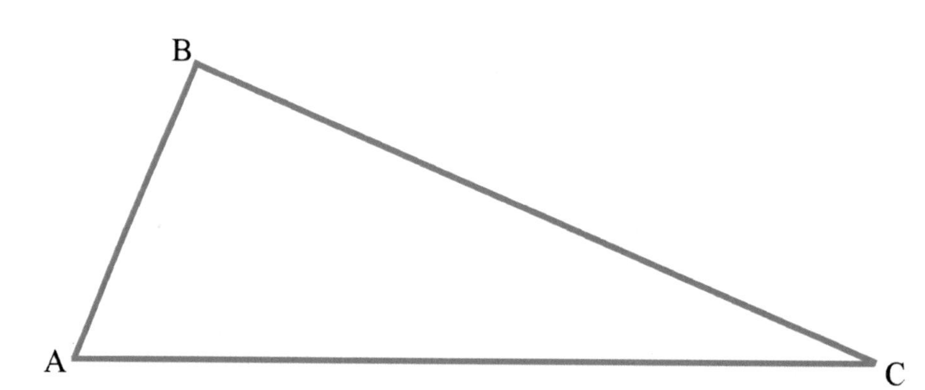

5. Measure some things in your classroom or at home *two times*. First measure them in inches, to the nearest half-inch. Them measure them in centimeters, to the nearest whole centimeter. Remember to write "about" if the thing is not exactly so many inches or centimeters. Write your results in the table below.

Item	in inches	in centimeters
	in.	cm
	in.	cm
	in.	cm
	in.	cm
	in.	cm

a. Which numbers are bigger, the centimeter-amounts or the inch-amounts?

b. Which measuring unit is bigger, one centimeter or one inch?

Notice: If your measuring unit is small (like 1 cm), you need MORE of them THAN if you use a longer measuring unit (inch).

c. Megan measured a spoon. It was 13 cm long. If she measures it in inches, will the result be more than 13 inches, or less than 13 in.?

d. Harry measured a toy car in inches. It was 3 in. If he measures it in centimeters, will the result be more than 3 cm, or less than 3 cm?

6. Draw three dots on a blank paper so you can join them and make a triangle. Then, measure its sides BOTH in inches (to the nearest half-inch) and in centimeters (to the nearest centimeter). Write your results in the table.

My Triangle	in inches	in centimeters
Side 1	in.	cm
Side 2	in.	cm
Side 3	in.	cm

How many centimeters is the *perimeter* (all the way around the shape)? __________ cm

How many inches is the *perimeter* (all the way around the shape)? __________ in.

Feet and Miles

This is a tape measure. The numbers 1, 2, 3, and so on, are inches.

Above number 12 you see "1F". That means *1 foot*. 12 inches equals 1 foot.

Unroll the tape measure some more, until you find "2 F" or "2 ft" (which means two feet), and "3 ft" (three feet), and so on. Stretch out the tape measure as far as you can. What is the most number of feet it has?

This tape measure has both inches and centimeters. The numbers on the top part are inches, and the numbers on the bottom part are centimeters. The number 60 means 60 cm, and the "1" after it means 61 cm.

You use *feet* as your measuring unit when you measure the width of a room or of a table, the length of a house, or of a swimming pool.

People often use both feet and inches. For example, a table can be 5 feet 10 inches long. Or, a boy can be 4 ft. 7 in. tall. How tall are you in feet and inches?

1. Use the tape measure to find distances in feet, or feet and inches. Let an adult help you.

Thing or distance	How long / tall
the room you are in	
a table	

2. How tall are these people? Ask your mom, dad, or others.

You: _____ ft. ______ in. ________________: _____ ft. ______ in.

Your mom: _____ ft. ______ in. ________________: _____ ft. ______ in.

3. Find three things you can measure in feet. But wait! First *guess* how long or tall they are. Then, check your guess by measuring.

Thing or distance	My guess	How long / tall

4. Now, measure again some of the things you already measured in feet, but this time measure them in centimeters. Or, you can still find new things to measure.

Thing or distance	centimeters	feet & inches

5. Which is a bigger (or longer) measuring unit, 1 centimeter or 1 foot?

Jared measured the height of a fridge twice, first in feet and then in centimeters.

It was 5 ft tall. How tall was it in centimeters? **a.** 15 cm **b.** 150 cm **c.** 3 cm

6. He also measured the height of a bucket twice, in feet and then in centimeters.

It was 60 cm tall. How tall was it in feet? **a.** 6 ft **b.** 100 ft **c.** 2 ft

7. Which is a longer measuring unit, a meter or a foot?

Jared measured the length of his room twice, first using feet and then using meters.

It was 4 m wide. How many feet wide was it? **a.** 2 ft **b.** 5 ft **c.** 12 ft

> Distances between towns or between countries are measured in *miles*.
> 1 mile is 5,280 feet (five-thousand two-hundred eighty)! That is a lot of feet—many, many more than your tape measure has.

8. Can you think of familiar distances in everyday life or in your neighborhood that are so many miles? An adult can help. You can also look in your social studies book.

Distance	How many miles

9. Aaron went on a trip with his family. On the first day, they drove 80 miles and visited a nature park. On the second day, they drove 200 miles. On the third day, they drove 110 miles back home.

   a. How long a distance did the family drive in all?

   b. How much longer distance did they drive on the second day than on the first day?

10. Which unit would you use to find the following distances: inches (in.), feet (ft), miles (mi), or feet and inches (ft. in.)?

Distance	Unit
from New York to Los Angeles	
from a house to a neighbor's	
the width of a notebook	
the distance around the earth	
how tall a refrigerator is	
the width of a porch	
the length of a board	

Distance	Unit
the length of a train	
the length of a playground	
from a train station to the next	
the width of a computer screen	

Meters and Kilometers

We use *meters* to measure medium and long distances.

Find a tape measure that has centimeters.
Find the 100th centimeter on it. That is the 1-meter point.

100 centimeters equals 1 meter.

1. **a.** Mark **one meter** on the floor. Can you take such a big step?
   Can the teacher?

   b. On the 1-metre line you marked, practice taking two steps that together are 1 meter long. Then take similar steps to *estimate* the length of a room (or if outside, a building).
   Count your steps: I took _______ steps
   Since you took 2 steps for each meter, find half of your count to get the length in meters. The room is about ______ m long.

   Measure to check your estimation.

   You can repeat this to estimate some other distance or length.

2. How tall are these people? Measure, or ask your mom, dad and friends.

   You: __________ cm   _________________: __________ cm

   Your mom: __________ cm   _________________: __________ cm

   Your dad: __________ cm   _________________: __________ cm

3. Measure some things using meters and centimeters. First guess how long or tall they are. Then check your guesses by measuring. Let an adult help you.

Item	My guess	How long/tall

Distances between towns or between countries are measured in _kilometers_.
1 kilometer is 1 000 meters (one thousand meters)!

4. Write in the table below **three distances** that are important in your life and are measured in kilometers. Ask an adult to help you. Examples include: from home to the library, from home to downtown, from home to Grandmother's, from your town to the capital city, etc.

From ... to	distance in km

5. The picture shows the field for Finnish baseball game ("pesäpallo"). How many meters do you run with these "routes"?

a. You run from the home base to the 1st base and then return to the home base.

b. You run from the home base to the 1st base and on to the 2nd base, plus one meter over, because you cannot stop in time.

c. (Challenge) You run all the way around the field.

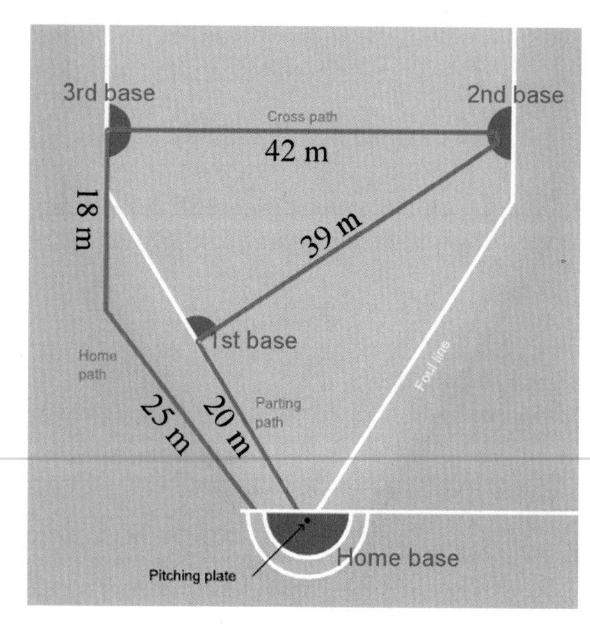

6. Which unit would you use to find these below: centimeters (cm), meters (m), or kilometers (km)?

Distance	Unit
the length of a park	
from Tshwane to the North Pole	
the length of a cell phone	
the length of a bus	

Distance	Unit
around your wrist	
the height of a room	
the length of an airplane trip	
the length of a grasshopper	

Weight in Pounds

Weight means *how heavy* something is. You can measure weight using a scale. A bathroom scale measures weight in *pounds* or in *kilograms*.

In this lesson you will need:

- a bathroom scale that measures in pounds
- a bucket and water
- encyclopedias or some other fairly heavy books
- a plastic bag or some other bag
- a backpack

The numbers on your scale may go up by twenties, and not by tens. In the picture here, the longer line halfway in-between the two numbers is TEN more than the smaller of the two numbers. Each little line means 2 pounds more than the previous line.

The scale on the right is stopped at the second little line after 140 pounds, which means 140 + 2 + 2 pounds, or 144 pounds.

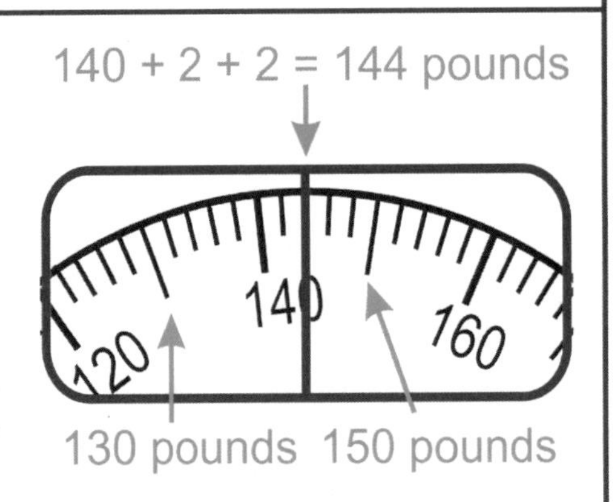

We use "lb" to abbreviate the word pounds. 15 pounds = 15 lb. The "lb" comes from the Latin word *libra*, which also means a pound.

1. How many pounds is the scale showing? You can mark the in-between ten-numbers on the scale to help.

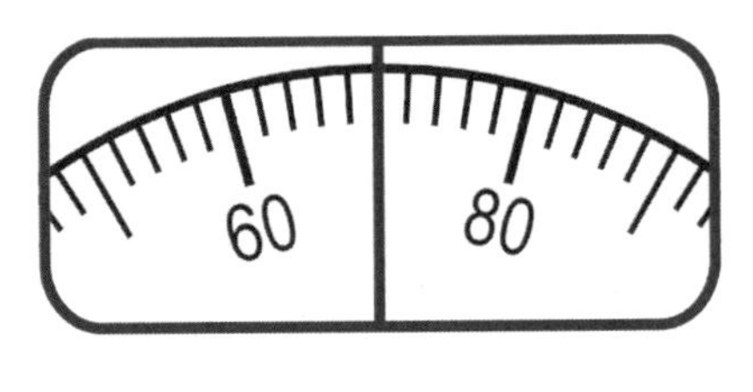

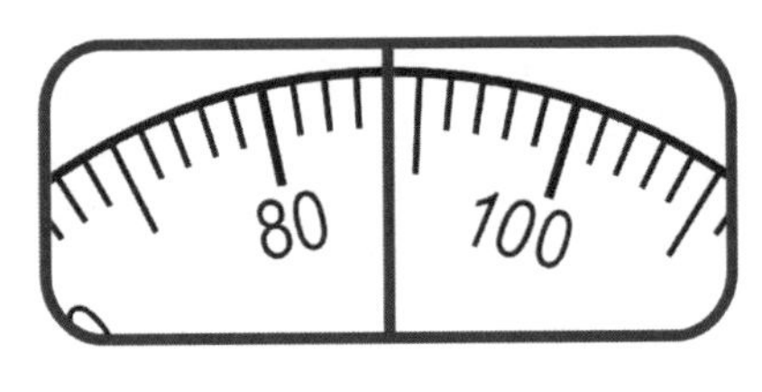

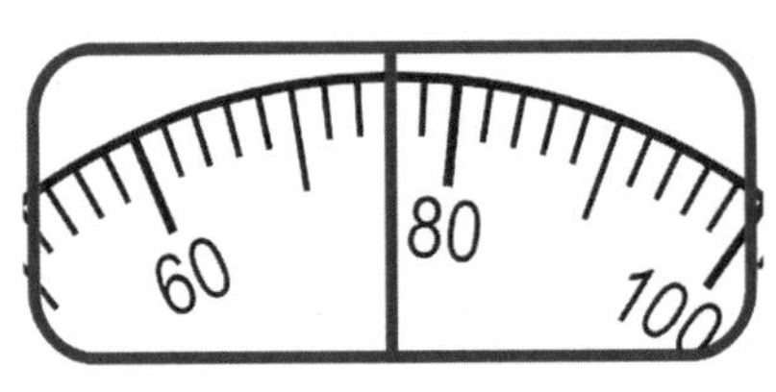

a. __________ lb **b.** __________ lb **c.** __________ lb

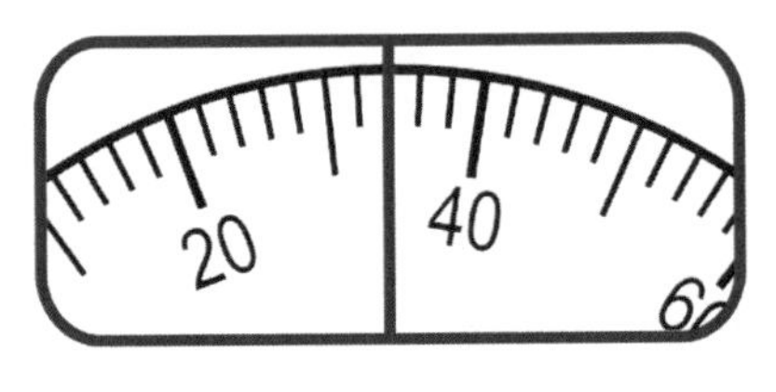

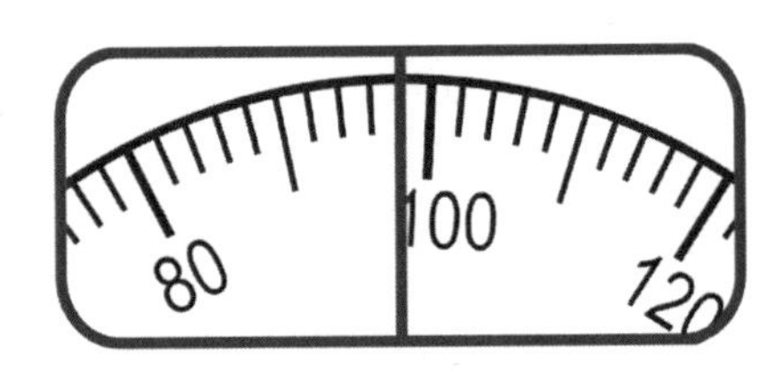

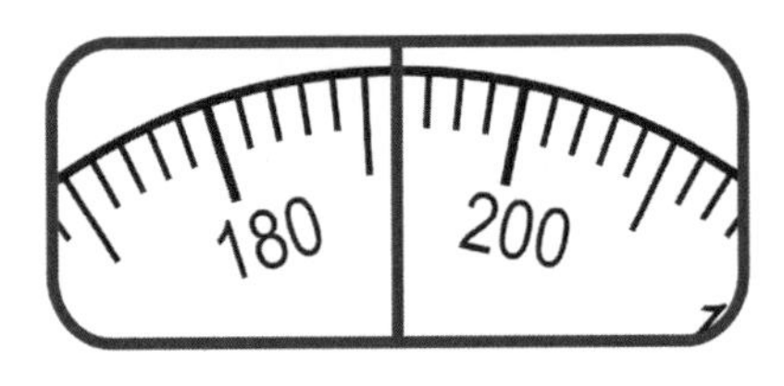

d. __________ lb **e.** __________ lb **f.** __________ lb

2. Step onto the scale. I weigh __________ pounds.

3. Find out how many pounds your family members weigh. Write a list below.

________________________ _____ lb ________________________ _____ lb

________________________ _____ lb ________________________ _____ lb

________________________ _____ lb ________________________ _____ lb

4. Weigh some other items. Note that on a bathroom scale, you cannot weigh very light items, nor very big and bulky ones because you cannot place them on the scales.

a bucket full of water _______ lb Mom's skillet _______ lb

a bucket half full of water _______ lb ________________________ _______ lb

a stack of heavy books _______ lb ________________________ _______ lb

5. Find out how many pounds of water you can carry. Can you carry the bucket when it is full? If not, pour out some water until you can carry the bucket.

I can carry a bucket of water that weighs _______ lb.

6. **a.** Find out how many pounds of books you can carry in a bag. Fill the bag with books and weigh it. Can you carry it? If not, take out some books until you are able to carry the bag.

I can carry a bagful of books that weighs _______ lb.

b. The same as above, but use a backpack. (Do you think you can carry more or less?)

I can carry a backpack that weighs _______ lb.

c. Weigh yourself with and without a heavy bag of books.

I weigh _______ lb. I weigh _______ lb with the heavy bag.

What is the difference? _______ lb.

d. Use the method above with a heavy book. The book weighs _______ lb.

Weight in Kilograms

Weight means <u>*how heavy*</u> something is. You can measure weight using *a scale*. A bathroom scale measures weight in *kilograms* (abbreviated kg).

The scale usually has short lines for each kilogram increment, and long lines for each 10 kilograms. In the picture below, the in-between numbers ending in "5" are marked with the number 5.

In this lesson, you need to use a bathroom scale that measures weight in kilograms. You will also need

- a bucket and water
- encyclopedias or some other fairly heavy books
- a plastic bag or some other bag
- a backpack

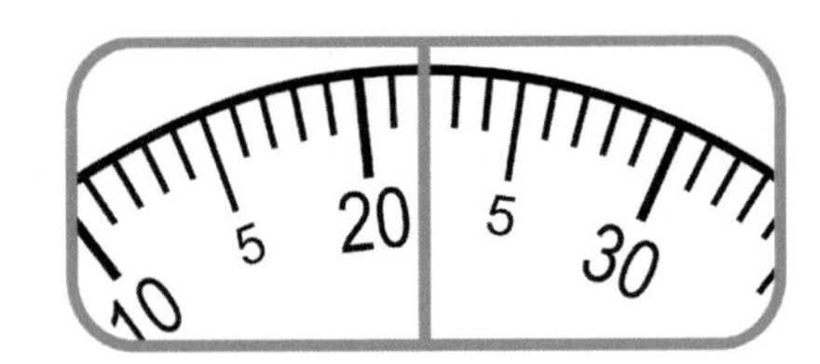

The scale is showing 22 kg.

1. How many kilograms is the scale showing?

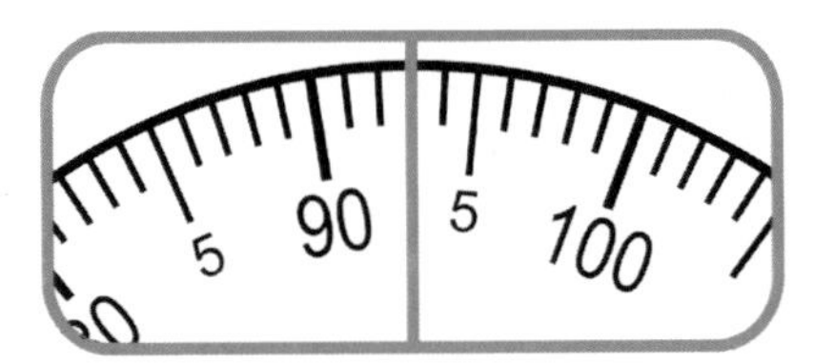

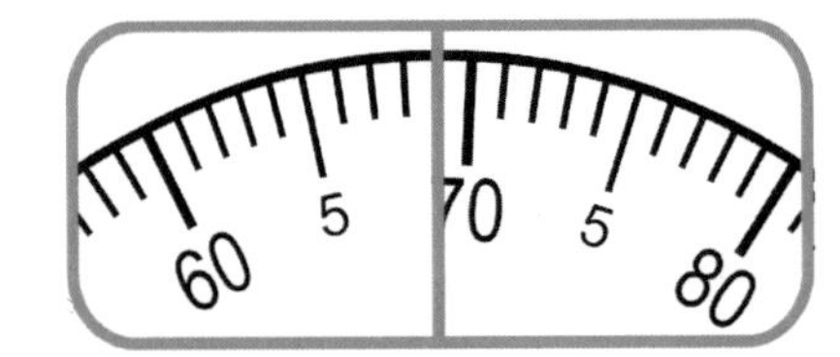

a. ____________________ **b.** ____________________ **c.** ____________________

2. Step onto the scale. How much do you weigh? _______ kg

3. Find out how many kilograms your family members weigh. Write a list below.

___________________________ _____ kg ___________________________ _____ kg

___________________________ _____ kg ___________________________ _____ kg

___________________________ _____ kg ___________________________ _____ kg

4. Weigh some of your family members together.

________________ and ____________________ together weigh ______ kg.

________________ and ____________________ together weigh ______ kg.

5. Now weigh some other items with the bathroom scale. Note: you cannot weigh very light items on it. You also cannot weigh very big and bulky items (such as tables) on it because you cannot place them fully on the scale. Try to find objects that are not very big.

a bucket full of water	_______ kg	Mom's frying pan	_______ kg
a bucket half full of water	_______ kg	___________________________	_______ kg
a stack of heavy books	_______ kg	___________________________	_______ kg

6. Find out how many kilograms of water you can carry. Can you carry the bucket when it is full? If not, pour out some water until you can carry the bucket.

   I can carry a bucket of water that weighs _______ kg.

7. **a.** Find out how many kilograms of books you can carry in a bag. Fill the bag with books and weigh it. Can you carry it? If not, take out some books until you are able to carry the bag.

   I can carry a bagful of books that weighs _______ kg.

   b. The same as above, but use a backpack.

   I can carry a backpack that weighs _______ kg.

   c. Weigh yourself with and without the heavy bag of books.

   I weigh _______ kg. I weigh _______ kg with the heavy bag.

   What is the difference? _______ kg.

   You can use this method to weigh items that cannot easily be placed on the scales, but that you can hold.

   d. Weigh yourself with and without a heavy book.

   I weigh _______ kg. I weigh _______ kg with the heavy book.

   What is the difference? _______ kg. So, the book weighs _______ kg.

Review

1. Which unit or units would you use for the following distances: inches (in.), feet (ft), miles (mi), centimeters (cm), or meters (m)? If two different units work, write both.

Distance	Unit or units
from your house to the grocery store	
from Miami to New York	
the distance across the room	
the height of a bookcase	

2. Measure this line to the nearest centimeter and to the nearest half-inch.

about _______ cm <u>or</u> about _________ in.

3. **a.** Draw a line that is 3 1/2 inches long.

b. Draw a line that is 9 cm long.

4. Measure these two pencils to the nearest centimeter, *and* to the nearest half-inch. Then fill in:

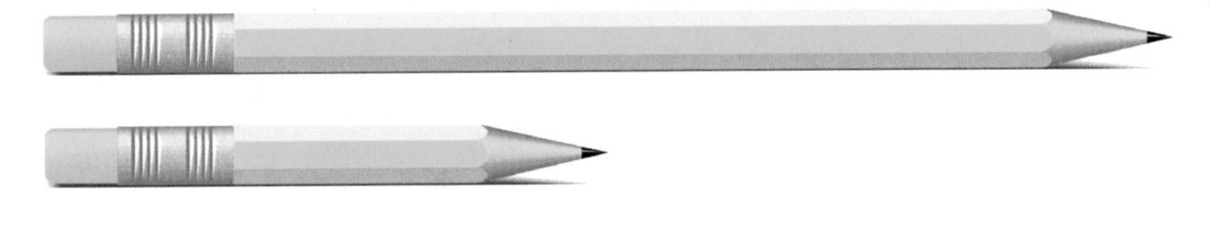

The longer pencil is about ________ cm longer than the shorter one.

The longer pencil is about ________ inches longer than the shorter one.

5. Measure the width and length of the room you are in. First, measure them using feet and inches. Then, measure them using meters and centimeters.

Width: _______ ft _______ in. *or* _______ m ________ cm

Length: _______ ft _______ in. *or* _______ m ________ cm

Measuring Grade 2 Answer Key

Measuring to the Nearest Centimeter, p. 9

1. a. 7 cm   b. 3 cm   c. 6 cm   d. 11 cm

2. a. 8 cm   b. 6 cm

3.

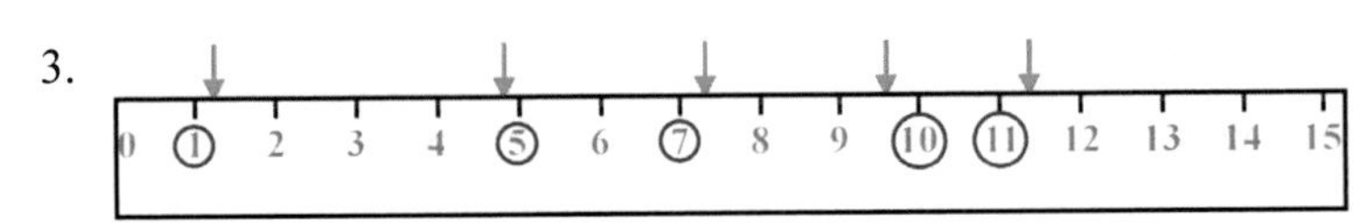

4. a. about 7 cm   b. about 4 cm   c. about 8 cm

5. a. about 6 cm   b. about 2 cm   c. about 4 cm

6. Answers will vary. Please check the students' work.

7. a.

b.

c.

8. Answers will vary.

Inches and Half-Inches, p. 12

1. a. 2 inches   b. 1 1/2 inches   c. 2 1/2 inches   d. 4 1/2 inches

2. a. 3 1/2 inches   b. 1 1/2 inches   c. 5 inches

3.

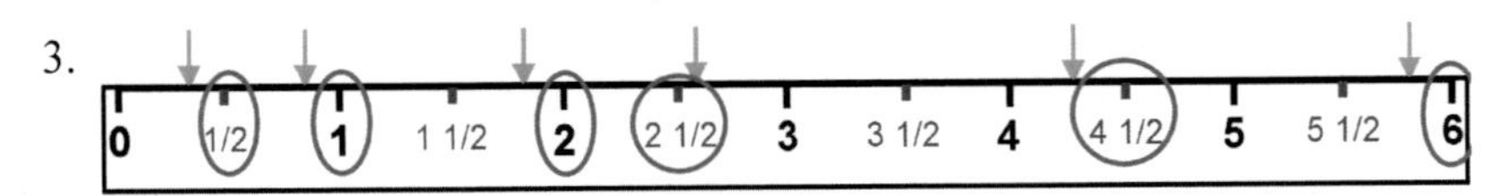

4. a. 3 inches   b. 3 1/2 inches   c. 4 1/2 inches   d. 5 inches

5. a 3 1/2 inches   b. 2 inches   c. 5 1/2 inches

6. Please check the student's work, as c. and d. are too long for the width of this page to show an example.

7. a. Rectangle, Side AB 2 1/2 inches, Side BC 1 inch, Side CD 2 1/2 inches, Side DA 1 inch.
All the way around 7 inches.
b. Square, Side AB 1 1/2 inches, Side BC 1 1/2 inches, Side CD 1 1/2 inches, Side DA 1 1/2 inches.
All the way around 6 inches.

Some More Measuring, p. 15

1. a. 2 pencils b. 4 pencils
   c. 4 pencils d. 3 pencils
   e. The shortest pencil is 3 cm and the longest pencil is 11 cm. The longest pencil is 8 cm longer than the shortest.

2. It is a quadrilateral. The perimeter is about 18 cm.

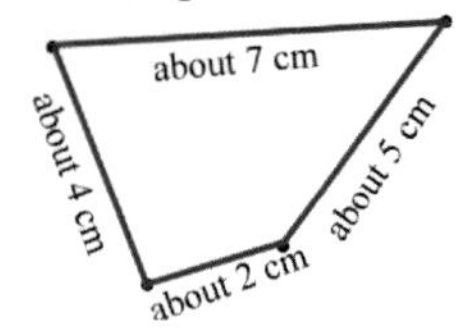

3. Answers will vary.

4. Side AB about 1 1/2 in. Side BC 4 in. Side CA 4 1/2 in. Perimeter about 10 in.

5. a. the centimeter-amounts b. one inch c. less than 13 inches d. more than 3 cm

6. Answers will vary.

Feet and Miles, p. 18

1.-4. Answers will vary.

5. 1 foot is longer than 1 cm. It was about 150 cm tall.

6. It was about 2 feet tall.

7. A meter is longer. About 12 feet.

8. Answers will vary.

9. a. 390 miles b. 120 miles

10.

Distance	Unit
from New York to Los Angeles	mi
from a house to a neighbor's house	ft
the width of a notebook	in.
the distance around the earth	mi
how tall a refrigerator is	ft. in.
the width of a porch	ft
the length of a board	ft. in.

Distance	Unit
the length of a train	ft
the length of a playground	ft
from a train station to the next	mi
the width of a computer screen	in.

Meters and Kilometers, p. 21

1. a. You might not be able to, but the teacher can.   b. Answers will vary.

2.-4. Answers will vary.

5. a. 20 m + 20 m = 40 m   b. 20 m + 39 m + 1 m = 60 m   c. 20 m + 39 m + 42 m + 18 m + 25 m = 144 m

6.

Distance	Unit
the length of a park	m
from Miami to the North Pole	km
the length of a cell phone	cm
the length of a bus	m & cm

Distance	Unit
around your wrist	cm
the height of a room	m & cm
the length of an airplane trip	km
the length of a grasshopper	cm

Weight in Pounds, p. 23

1. a.70 pounds   b. 88 pounds   c. 76 pounds   d. 34 pounds   e. 98 pounds   f. 192 pounds

2-6. Answers will vary.

Weight in Kilograms, p. 25

1. a. 45 kilograms   b. 93 kilograms   c. 69 kilograms

2-7. Answers will vary.

Review, p. 27

1.

Distance	Unit or units
from your house to the grocery store	mi or km
from Miami to New York	mi or km
the distance across the room	m or ft
the height of a bookcase	ft, in, m, or cm

2. About 6 cm *or* about 2 1/2 in.

3. a.

   b.

4. The longer pencil is about 7 cm longer than the shorter one.
   The longer pencil is about 2.5 inches longer than the shorter one.

5. Answers will vary.

Appendix: Common Core Alignment

The table below lists each lesson and next to it the relevant Common Core Standard.

Lesson	page number	Standards
Measuring to the Nearest Centimeter	9	2.MD.1 2.MD.3 2.MD.4
Inches and Half-Inches	12	2.MD.1 2.MD.3
Some More Measuring	15	2.MD.1 2.MD.2 2.MD.4 2.MD.9 2.G.1
Feet and Miles	18	2.MD.1 2.MD.2 2.MD.3 2.MD.5
Meters and Kilometers	21	2.MD.1 2.MD.3 2.MD.5
Weight in Pounds	23	3.MD.2
Weight in Kilograms	25	3.MD.2
Review	27	2.MD.1 2.MD.2 2.MD.4

35840070R00020

Made in the USA
Middletown, DE
17 October 2016